Microbiology Demystified: A Student's Guide to Understanding the Invisible World

SUESS

Microbiology Demystified: A Student's Guide to Understanding the Invisible World

Copyright © 2023 by SUESS

All rights reserved. No part of this book may be reproduced or transmitted in any form or by any means, electronic or mechanical, including photocopying, recording, or by any information storage and retrieval system, without permission in writing from the publisher.

This book is a work of fiction. Names, characters, places, and incidents either are the product of the author's imagination or are used fictitiously. Any resemblance to actual events, locales, persons, living or dead, is entirely coincidental.

The first edition was published in 2023

ISBN:
Published by:
Sunshine
1663 Liberty Drive
Hyderabad, IN 47403
www.Sunshinepublishers.com

This book is self-published using on-demand printing and publishing, which allows it to be printed and distributed globally

TABLE OOF CONTENT

Chapter 1: Introduction to Microbiology

The Importance of Microbiology in Our Lives

Microbiology, the study of microscopic organisms, plays a vital role in our everyday lives. From understanding the causes of diseases to the development of life-saving medications, the field of microbiology has revolutionized our understanding of the invisible world that surrounds us. In this subchapter, we will explore the significance of microbiology in our lives, particularly for students with an interest in biology.

One of the primary contributions of microbiology is its role in the field of medicine. Through the study of microorganisms, scientists and healthcare professionals have been able to identify and understand the causes of various diseases. This knowledge has led to the development of effective treatments and vaccines, saving countless lives. For biology students, understanding microbiology is essential for a comprehensive understanding of human health and the mechanisms of diseases.

Microbiology also plays a crucial part in the field of biotechnology. By harnessing the power of microorganisms, scientists can produce valuable products such as antibiotics, enzymes, and renewable energy sources. Furthermore, microbiology is at the forefront of genetic engineering and the development of genetically modified organisms (GMOs). Students interested in biology can explore these exciting applications of microbiology, which have the potential to shape the future of medicine, agriculture, and environmental sustainability.

Moreover, microbiology is essential for understanding the natural world around us. Microorganisms are ubiquitous and play essential roles in various ecosystems. They are involved in nutrient cycling, decomposition, and symbiotic relationships with plants and animals. By studying microbiology, students can gain insights into the intricate web of life and how microorganisms contribute to the overall health and balance of our planet.

Lastly, microbiology is a field that offers numerous career opportunities. With advancements in technology and research, the demand for microbiologists is steadily increasing. Graduates with a background in microbiology can pursue careers in healthcare, pharmaceuticals, environmental sciences, food safety, and research. By delving into the world of microbiology, students can open doors to exciting and fulfilling professional paths.

In conclusion, microbiology is a discipline that holds immense importance in our lives. From its contributions to medicine and biotechnology to its role in understanding the natural world, microbiology offers students of biology a captivating and dynamic field of study. By exploring the invisible world of microorganisms, students can gain a deeper understanding of life and its intricate processes, paving the way for a brighter and healthier future.

The History and Milestones of Microbiology

Microbiology is a fascinating field that explores the invisible world of microorganisms. In this subchapter, we will delve into the history and milestones of microbiology, tracing its origins and significant contributions to the field of biology.

The story of microbiology begins in the late 17th century with the invention of the microscope. This groundbreaking invention allowed scientists to observe and study organisms that were too small to be seen by the naked eye. Antonie van Leeuwenhoek, a Dutch scientist, was one of the pioneers in this field. He made significant contributions by documenting his observations of various microorganisms, including bacteria and protozoa.

In the 19th century, the field of microbiology gained momentum with the work of Louis Pasteur. Pasteur's experiments on fermentation and the role of microorganisms in causing diseases revolutionized the field. He also developed the process of pasteurization, which involves heating liquids to kill bacteria and other harmful microbes.

Another milestone in the history of microbiology was the development of the germ theory of disease. This theory, proposed by Robert Koch, stated that specific microorganisms cause specific diseases. Koch's postulates provided a framework for identifying and studying disease-causing microorganisms, leading to significant advancements in the understanding and treatment of infectious diseases.

The discovery of antibiotics in the early 20th century was yet another landmark in microbiology. Alexander Fleming's accidental discovery of penicillin paved the way for the development of numerous antibiotics that have saved millions of lives. This discovery revolutionized the field of medicine and brought microbiology to the forefront of scientific research.

Advancements in technology, such as the invention of electron microscopes and DNA sequencing techniques, have further propelled the field of microbiology. These tools have enabled scientists to study microorganisms at a molecular level, unraveling their genetic makeup and understanding their complex interactions with the environment and other organisms.

Today, microbiology continues to be a vibrant and rapidly evolving field. It plays a crucial role in various disciplines, including medicine, agriculture, and environmental science. Microbiologists are at the forefront of battling infectious diseases, developing new treatments, and exploring the vast diversity of microorganisms that inhabit our planet.

As students of biology, understanding the history and milestones of microbiology is essential. It provides a foundation for comprehending the immense impact that microorganisms have on our lives and the interconnectedness of all living things. By studying the invisible world of microorganisms, we can unlock the secrets of life itself and contribute to the advancement of scientific knowledge.

The Scope and Branches of Microbiology

Microbiology is the scientific study of microorganisms, which are microscopic organisms such as bacteria, viruses, fungi, and protozoa. These tiny organisms are invisible to the naked eye, yet they play a crucial role in shaping the world we live in. In this subchapter, we will explore the scope and branches of microbiology, shedding light on the vast and fascinating world of these invisible creatures.

The scope of microbiology is vast, encompassing a wide range of disciplines and applications. Microbiologists study the diversity, structure, function, and interactions of microorganisms, as well as their impact on human health, ecology, industry, and agriculture. Understanding microorganisms is crucial for various fields, including medicine, biotechnology, environmental science, and food production.

The branches of microbiology are divided into several subdisciplines, each focusing on different aspects of microbial life. One of the key branches is medical microbiology, which involves the study of microorganisms that cause diseases in humans. By identifying and understanding these disease-causing microorganisms, medical microbiologists play a crucial role in diagnosing and treating infections.

Another important branch is environmental microbiology, which focuses on the role of microorganisms in the environment. Environmental microbiologists study how microorganisms impact ecosystems, help recycle nutrients, and play a role in bioremediation, which is the use of microorganisms to clean up polluted environments.

Industrial microbiology is yet another branch that examines the use of microorganisms in various industries such as food and beverage, pharmaceuticals, and biotechnology. Microbes are used to produce antibiotics, enzymes, biofuels, and other valuable products.

Microbial genetics and molecular biology explore the genetic makeup and molecular mechanisms of microorganisms. These branches help scientists understand how microorganisms evolve, how they adapt to different environments, and how they interact with other organisms.

Food microbiology focuses on the study of microorganisms that affect food quality and safety. Microbiologists in this field work to prevent foodborne illnesses by identifying and controlling harmful microorganisms in the food production process.

These are just a few examples of the branches of microbiology, highlighting the diverse areas where microorganisms have a significant impact. As students of biology, understanding the scope and branches of microbiology is essential to appreciate the vastness and importance of the invisible world that surrounds us. With this knowledge, you will be equipped to explore the depths of microbiology and contribute to advancements in this exciting field.

Chapter 2: Microbial Structure and Function

Prokaryotic vs. Eukaryotic Cells

In the fascinating world of microbiology, understanding the differences between prokaryotic and eukaryotic cells is essential. These two types of cells form the basis of all living organisms, and comprehending their distinctions is crucial for students studying biology.

Prokaryotic cells are the simplest and most ancient form of life on Earth. They lack a nucleus, which means their genetic material, or DNA, is not enclosed within a membrane. Instead, prokaryotes have a single, circular chromosome that floats freely in the cytoplasm. Bacteria and archaea are examples of prokaryotic organisms. These cells also lack membrane-bound organelles, such as mitochondria or chloroplasts, which are essential for energy production and photosynthesis. Instead, prokaryotes generate energy through their cell membrane.

On the other hand, eukaryotic cells are more complex and evolved. They possess a true nucleus, which houses their DNA. This nucleus is surrounded by a double membrane, providing protection and organization for the genetic material. Eukaryotes can be found in various kingdoms, including animals, plants, fungi, and protists. These cells contain membrane-bound organelles that perform specific functions, such as mitochondria, which produce energy, and chloroplasts, responsible for photosynthesis in plants.

One notable difference between prokaryotes and eukaryotes is their size. Prokaryotic cells are generally smaller, ranging from 0.1 to 5 micrometers, while eukaryotic cells are larger, typically measuring 10 to 100 micrometers. This size difference allows eukaryotes to accommodate more complex internal structures and organelles.

Another significant distinction lies in their reproduction methods. Prokaryotes reproduce asexually through binary fission, where the parent cell divides into two identical daughter cells. Eukaryotes, on the other hand, undergo cell division through a process called mitosis, which ensures the equal distribution of genetic material between daughter cells.

Understanding these differences is crucial for various branches of biology, including genetics, evolution, and medical research. By studying prokaryotic and eukaryotic cells, students can gain insights into the fundamental processes of life and explore the intricacies of microorganisms. These insights are key to understanding diseases caused by bacteria or fungi and developing new treatments to combat them.

In conclusion, the differences between prokaryotic and eukaryotic cells are vital knowledge for students of biology. From their structure to their reproductive methods, each cell type has unique characteristics that shape the invisible world of microbiology. By unraveling the mysteries of these cells, students can gain a deeper understanding of the invisible world and contribute to advancements in the field of microbiology.

Cell Wall Structures and Functions

One of the fundamental components of a bacterial cell is its cell wall. The cell wall serves as a protective layer surrounding the cell, giving it shape and protection against external threats. In this subchapter, we will explore the structures and functions of cell walls, focusing on their significance in the field of microbiology.

The cell wall of bacteria is primarily composed of a unique molecule called peptidoglycan. Peptidoglycan consists of long chains of sugar molecules that are cross-linked by short peptides. This intricate network provides strength and rigidity to the cell wall, allowing bacteria to maintain their shape even in challenging environments.

The primary function of the cell wall is to protect the bacterial cell from osmotic pressure. Without a cell wall, bacteria would be susceptible to bursting due to the influx of water. The peptidoglycan layer prevents this by maintaining the structural integrity of the cell, allowing it to withstand changes in osmotic pressure.

Additionally, the cell wall plays a crucial role in cell division. As bacteria reproduce, the cell wall expands and eventually splits into two, ensuring the formation of two genetically identical daughter cells. Understanding the intricacies of cell wall synthesis is essential in combating bacterial infections, as disrupting this process can hinder bacterial growth and reproduction.

The composition of the cell wall also influences the staining properties of bacteria. This property is extensively utilized in microbiological techniques, such as Gram staining, which helps classify bacteria into two major groups: Gram-positive and Gram-negative. Gram-positive bacteria have a thick peptidoglycan layer, retaining the crystal violet stain, while Gram-negative bacteria have a thinner peptidoglycan layer and take up the counterstain, safranin.

Lastly, the cell wall acts as a barrier against the entry of certain molecules, including antibiotics. By understanding the structure and composition of the cell wall, scientists can develop strategies to overcome this barrier and improve the efficacy of antibiotic treatments.

In conclusion, the cell wall is a vital component of bacterial cells, providing structural support, protection against osmotic pressure, and playing a critical role in cell division. Its composition and properties have profound implications in microbiology, aiding in bacterial classification, understanding antibiotic resistance, and the development of therapeutic interventions. By delving into the intricacies of cell wall structures and functions, students can gain a deeper understanding of the invisible world of microbiology.

Cell Membrane and Transport Mechanisms

The cell membrane is a vital structure that encloses all living cells, serving as a protective barrier and regulating the transport of molecules in and out of the cell. Understanding the cell membrane and its transport mechanisms is fundamental in the field of biology, as it forms the basis of all cellular processes.

The cell membrane is composed of a phospholipid bilayer, which consists of two layers of phospholipids arranged in such a way that their hydrophobic tails are facing each other, while their hydrophilic heads are exposed to the surrounding environment. This unique structure allows the cell membrane to be selectively permeable, meaning that it only allows certain molecules to pass through while blocking others.

Transport mechanisms are the processes by which molecules move across the cell membrane. There are two main types of transport mechanisms: passive transport and active transport.

Passive transport is the movement of molecules across the cell membrane without the input of energy. It can occur through two processes: diffusion and osmosis. Diffusion is the movement of molecules from an area of high concentration to an area of low concentration, driven by the natural tendency of molecules to spread out and reach equilibrium. Osmosis, on the other hand, is the diffusion of water molecules across a selectively permeable membrane.

Active transport, on the other hand, requires the input of energy to move molecules against their concentration gradient, from an area of low concentration to an area of high concentration. This process is essential for the uptake of nutrients and the removal of waste products from the cell.

In addition to diffusion, osmosis, and active transport, there are also other specialized transport mechanisms such as facilitated diffusion and endocytosis/exocytosis. Facilitated diffusion involves the use of transport proteins to facilitate the movement of specific molecules across the cell membrane. Endocytosis and exocytosis, on the other hand, are processes by which large molecules or particles are transported into or out of the cell by the formation of vesicles.

Understanding the cell membrane and its transport mechanisms is crucial in biology as it allows us to comprehend how cells maintain their internal environment, regulate the transport of essential molecules, and communicate with other cells. By studying these mechanisms, students can gain insights into the intricate workings of cells and their role in various biological processes.

Organelles and their Roles in Microbial Cells

In the microscopic world of microorganisms, such as bacteria and archaea, there exists a fascinating array of structures called organelles, which play crucial roles in the survival and function of these microbial cells. Understanding these organelles and their functions is essential for students delving into the realm of microbiology.

One of the most prominent organelles in microbial cells is the cell membrane, also known as the plasma membrane. This thin layer regulates the flow of molecules into and out of the cell, maintaining its internal environment. It acts as a barrier, protecting the cell from harmful substances in the external environment while allowing necessary nutrients to enter. Additionally, the cell membrane houses proteins that aid in cell signaling and communication.

Another vital organelle found in microbial cells is the cytoplasm, a gel-like substance that fills the cell. It contains various structures, including ribosomes, which are responsible for protein synthesis. These tiny organelles read the genetic information from the DNA to produce proteins that carry out essential cellular functions.

Microbial cells also possess a distinct structure called the nucleoid, which is not enclosed by a membrane. The nucleoid is where the microbial cell stores its genetic material, usually in the form of a circular DNA molecule called a chromosome. This genetic material contains the instructions for the cell's growth, reproduction, and overall functioning.

Some microbial cells possess additional organelles called flagella and pili. Flagella are whip-like structures that rotate to propel the cell through fluids, allowing it to move towards beneficial environments or escape from harmful ones. Pili, on the other hand, are hair-like structures that enable microbial cells to attach to surfaces, including host tissues, aiding in infection and colonization.

In certain microbial cells, specialized organelles called mitochondria are present. Mitochondria are responsible for generating energy in the form of adenosine triphosphate (ATP) through a process called cellular respiration. This energy is crucial for microbial cells to carry out their metabolic activities.

These are just a few examples of the organelles found in microbial cells and their roles in enabling these organisms to survive and thrive. Understanding these structures and their functions is essential for students studying microbiology, as it provides insight into the intricate world of microorganisms and their interactions with the environment.

By comprehending the roles of organelles in microbial cells, students can gain a deeper understanding of the invisible world of microorganisms, paving the way for advancements in various branches of biology, such as medical microbiology, environmental microbiology, and biotechnology.

Chapter 3: Microbial Growth and Reproduction

Nutritional Requirements for Microbial Growth

In the vast and diverse microbial world, organisms have different nutritional requirements for their growth and survival. Understanding these requirements is crucial in studying the invisible world of microbiology. In this subchapter, we will explore the essential nutritional needs of microorganisms and how they contribute to their growth and survival.

Microbes, just like any other living organisms, require specific nutrients to fulfill their metabolic needs. These nutrients can be divided into two main categories: macronutrients and micronutrients. Macronutrients, including carbon, nitrogen, phosphorus, sulfur, and oxygen, are needed in larger quantities, while micronutrients, such as trace metals like iron, copper, and zinc, are essential in smaller quantities.

Carbon is one of the fundamental macronutrients required by microorganisms. It serves as the structural backbone for all organic molecules and is obtained by microorganisms in various forms, such as sugars, amino acids, and organic acids. Nitrogen is another essential macronutrient, necessary for the synthesis of proteins and nucleic acids. Microbes can acquire nitrogen from organic sources like amino acids or inorganic compounds like ammonia or nitrate.

Phosphorus is crucial for the synthesis of nucleic acids and ATP, the energy currency of the cell. Microorganisms obtain phosphorus in the form of phosphate, which is readily available in the environment. Sulfur is necessary for the synthesis of certain amino acids and vitamins. It is obtained by microbes from sulfate or organic sulfur compounds.

Apart from these macronutrients, microorganisms also require smaller quantities of micronutrients, also known as trace elements. These elements play crucial roles as cofactors in enzyme reactions and are essential for the proper functioning of microbial cells. Iron, for example, is required for electron transport chains and enzymatic reactions involved in cellular respiration.

Understanding the nutritional requirements of microorganisms is not only important for their growth and survival but also has practical applications in various fields. For example, knowing the specific nutrient requirements of pathogenic bacteria can aid in the development of targeted antimicrobial agents and therapies.

In conclusion, the nutritional requirements for microbial growth are diverse and essential for their survival. Microbes require macronutrients like carbon, nitrogen, phosphorus, sulfur, and oxygen, as well as micronutrients like trace metals. Understanding these requirements helps us comprehend the invisible world of microorganisms and their impact on biology.

Physical Factors Affecting Microbial Growth

In the realm of microbiology, understanding the physical factors that influence microbial growth is essential. Microorganisms, including bacteria, fungi, and viruses, have specific requirements for their growth and survival. These factors vary widely and play a crucial role in determining the abundance and distribution of microbes in different environments. This subchapter will delve into the key physical factors that affect microbial growth, shedding light on their significance in the invisible world of microbiology.

Temperature is one of the most critical physical factors influencing microbial growth. Microbes exhibit distinct temperature preferences, falling into three categories: psychrophiles (thriving in cold environments), mesophiles (preferring moderate temperatures), and thermophiles (thriving in high temperatures). Understanding these temperature ranges is vital in predicting where certain microbes are likely to flourish.

Another important physical factor is pH, which refers to the acidity or alkalinity of a medium. Microbes have specific pH requirements, with some thriving in acidic conditions (acidophiles), while others prefer alkaline environments (alkaliphiles). Studying the pH preferences of microbes is crucial in various fields, such as food preservation and wastewater treatment.

Moisture content also plays a significant role in microbial growth. Microbes require a certain level of water activity to survive and reproduce. Understanding the moisture requirements of different microbes is essential in areas such as food storage and preservation, where controlling moisture content can prevent spoilage and microbial contamination.

Oxygen availability is another crucial physical factor affecting microbial growth. While some microbes require oxygen to grow (obligate aerobes), others cannot tolerate it and grow in its absence (obligate anaerobes). Additionally, some microbes can grow in the presence or absence of oxygen (facultative anaerobes). Understanding the oxygen preferences of different microbes is vital in areas such as medical microbiology and environmental studies.

Light is a physical factor that affects the growth of certain microbes, particularly photosynthetic organisms such as algae and cyanobacteria. These organisms utilize light energy to produce organic compounds through photosynthesis. The intensity, duration, and quality of light are crucial factors influencing the growth and distribution of photosynthetic microbes.

In conclusion, understanding the physical factors that affect microbial growth is vital in the field of microbiology. Temperature, pH, moisture content, oxygen availability, and light are among the significant physical factors that influence the abundance and distribution of microbes. Understanding these factors enables scientists to predict microbial growth patterns, control microbial populations, and apply microbiological knowledge to various fields such as medicine, agriculture, and environmental science. By delving into the invisible world of microbes, students of biology can gain a deeper understanding of the intricate interactions between microorganisms and their physical surroundings.

The Cell Cycle and Reproduction in Microorganisms

Microorganisms, the tiny living organisms that are invisible to the naked eye, play a vital role in the invisible world of microbiology. These microorganisms, whether they are bacteria, fungi, or protozoa, undergo a process known as the cell cycle to reproduce and perpetuate their species. Understanding the cell cycle in microorganisms is crucial for biologists and students studying the fascinating field of microbiology.

The cell cycle is a series of events that takes place in a cell, leading to its division and the formation of new cells. In microorganisms, this process is relatively simple compared to higher organisms like plants and animals. The cell cycle in microorganisms consists of four main stages: growth, DNA replication, division, and cytokinesis.

The first stage, known as growth, is the period when the microorganism prepares itself for replication. During this stage, the cell increases in size, synthesizing the necessary proteins, enzymes, and other molecules required for replication. Once the cell has reached its optimal size, it moves into the next stage.

The DNA replication stage is a crucial step in the cell cycle, where the microorganism duplicates its genetic material. DNA, the blueprint of life, contains all the instructions needed for the microorganism's survival and reproduction. The replication process ensures that each new cell receives an exact copy of the genetic material, ensuring genetic continuity.

After DNA replication, the cell enters the division stage. Here, the microorganism divides its cytoplasm and genetic material into two equal parts, forming two daughter cells. This division process can occur through various mechanisms, such as binary fission in bacteria or budding in yeast. The method of division depends on the type of microorganism and its specific characteristics.

Finally, the cell cycle concludes with cytokinesis, the physical separation of the two daughter cells. During cytokinesis, the microorganism divides its cytoplasm, membrane, and other cellular components, ensuring that each daughter cell becomes an independent and functional organism.

Understanding the cell cycle in microorganisms is crucial for several reasons. Firstly, it provides insights into the fundamental processes of life and reproduction. Secondly, it helps scientists understand the growth and proliferation of harmful microorganisms, contributing to the development of strategies to combat infectious diseases. Finally, studying the cell cycle in microorganisms can shed light on the evolution and diversity of life on Earth.

In conclusion, the cell cycle and reproduction in microorganisms are essential concepts in microbiology. This subchapter has provided an overview of the cell cycle, highlighting its stages and significance. By delving into the invisible world of microorganisms, students studying biology can gain a deeper understanding of the intricacies of life and the invisible organisms that surround us.

Control of Microbial Growth: Antimicrobial Agents

In the world of microbiology, understanding the mechanisms and applications of antimicrobial agents is crucial. This subchapter delves into the topic of controlling microbial growth through the use of these agents, providing students with a comprehensive guide to the invisible world of microorganisms.

Antimicrobial agents are substances that can inhibit or kill microorganisms, including bacteria, viruses, fungi, and parasites. They are essential tools in various fields, including medicine, agriculture, and food preservation. By understanding how these agents work, students can gain insights into their importance in combating infectious diseases and maintaining public health.

The subchapter begins by introducing the different types of antimicrobial agents, including antibiotics, antiviral drugs, antifungal agents, and antiparasitic drugs. Students will explore the mechanisms of action for each type, focusing on how they target specific components or processes within microorganisms. This understanding is vital for comprehending the importance of selecting the appropriate agent for a given infection.

The subchapter also covers the concept of drug resistance, a growing concern in the field of microbiology. Students will learn about the factors contributing to the emergence and spread of drug-resistant microorganisms, such as misuse and overuse of antimicrobial agents. They will gain insights into the importance of responsible antibiotic usage and the development of new strategies to combat drug resistance.

Furthermore, the subchapter explores the challenges in developing new antimicrobial agents. Students will discover the complex process of drug discovery and the obstacles faced by scientists in finding effective treatments for emerging infections. They will also learn about the role of research and innovation in overcoming these challenges and improving patient outcomes.

To enhance students' understanding, real-world examples and case studies are incorporated throughout the subchapter. These examples highlight the successful use of antimicrobial agents in treating specific infections or outbreaks, showcasing the significant impact these agents have on human health.

Overall, this subchapter on the control of microbial growth through antimicrobial agents provides students in the field of biology with a valuable resource. By exploring the mechanisms, applications, and challenges associated with these agents, students will gain a deeper understanding of their role in combating infectious diseases and protecting public health in the invisible world of microorganisms.

Chapter 4: Microbial Diversity and Classification

Taxonomy and Classification of Microorganisms

In the vast and diverse world of microorganisms, taxonomy and classification play a crucial role in understanding and organizing these invisible creatures. Taxonomy is the science of classifying living organisms, while classification is the process of categorizing them into groups based on their characteristics and evolutionary relationships. By studying the taxonomy and classification of microorganisms, students can gain a deeper understanding of their diversity and importance in various biological processes.

Microorganisms are microscopic organisms that include bacteria, archaea, fungi, protozoa, and viruses. Each group possesses unique characteristics that differentiate them from one another. Bacteria are single-celled organisms that lack a nucleus, while archaea are similar but have distinct genetic and biochemical features. Fungi, on the other hand, are eukaryotic organisms that possess a nucleus and play vital roles in decomposition and nutrient recycling. Protozoa are unicellular eukaryotes that are often found in aquatic environments and can be parasitic or free-living. Lastly, viruses are non-living entities that require a host cell for replication.

To classify microorganisms, scientists use a hierarchical system called the Linnaean system of taxonomy. This system involves organizing organisms into a hierarchy of groups, starting from the broadest category (domain) down to the smallest (species). The three domains of life are Bacteria, Archaea, and Eukarya. Each domain is further divided into various kingdoms, such as the Kingdom Protista, Kingdom Fungi, and Kingdom Plantae.

Within each kingdom, microorganisms are classified based on their shared characteristics, such as cell structure, mode of reproduction, and metabolic capabilities. For instance, bacteria can be further classified into different phyla, classes, orders, families, genera, and species. The classification system allows scientists to identify and name microorganisms, facilitating the study of their distribution, evolution, and ecological roles.

Understanding the taxonomy and classification of microorganisms has significant implications in various fields, including medicine, agriculture, and environmental science. It helps scientists identify disease-causing microorganisms and develop targeted treatments. In agriculture, it aids in the study of beneficial microorganisms that enhance plant growth and protect against pests. Additionally, studying the diversity of microorganisms in different environments contributes to our understanding of ecological processes and the preservation of ecosystems.

In conclusion, taxonomy and classification are essential tools for studying microorganisms in the field of biology. By organizing and categorizing these invisible creatures, scientists can better comprehend their diversity, evolutionary relationships, and ecological roles. A solid understanding of taxonomy and classification provides students with a foundation for further exploration of the fascinating world of microorganisms and their impact on the natural world.

Bacteria: Morphology, Identification, and Classification

In the fascinating world of microbiology, bacteria hold a significant place. These tiny microorganisms are found everywhere, from the depths of the ocean to the soil beneath our feet. Understanding the morphology, identification, and classification of bacteria is crucial for students of biology who wish to comprehend the invisible world that surrounds us.

Morphology refers to the study of the physical characteristics of bacteria. Despite their minuscule size, bacteria exhibit a diverse range of shapes and structures. Some bacteria are spherical and are called cocci, while others are rod-shaped and referred to as bacilli. There are also spiral-shaped bacteria known as spirilla or spirochetes. Understanding the morphology of bacteria can provide important clues about their behavior and potential impact on the environment and human health.

Identification of bacteria involves different techniques that allow scientists to determine the specific species or strain. One common method is staining, where bacteria are treated with a dye that highlights their cellular structures. Gram staining, for example, can distinguish between two major groups of bacteria: Gram-positive and Gram-negative. Another approach is the use of selective media, which contain specific nutrients that only allow certain bacteria to grow. These techniques, among others, are essential tools in the identification of bacteria and help researchers and medical professionals in diagnosing and treating bacterial infections.

Classification of bacteria involves grouping them into categories based on shared characteristics. The most common system used is the Bergey's Manual of Systematic Bacteriology, which classifies bacteria based on their genetic relationships, metabolism, and other factors. Bacteria are divided into phyla, classes, orders, families, genera, and species, allowing scientists to organize and study these microorganisms systematically.

Understanding the morphology, identification, and classification of bacteria is crucial for various fields. In medicine, it helps in diagnosing bacterial infections and developing effective treatments. In environmental science, it aids in studying the role of bacteria in ecosystems and their potential for bioremediation. In agriculture, it assists in identifying beneficial bacteria for crop enhancement and controlling plant diseases.

As students of biology, delving into the world of bacteria's morphology, identification, and classification opens up a realm of possibilities for further research and understanding. By comprehending these fundamental aspects, students can gain a deeper knowledge of the invisible world that shapes our lives and contributes to the intricate web of life on Earth.

Archaea: Unique Characteristics and Classification

In the vast world of microbiology, one group of organisms stands out for its unique characteristics and classification - the Archaea. Archaea are a fascinating group of microorganisms that differ from both bacteria and eukaryotes in several ways. In this subchapter, we will explore the distinctive features and classification of Archaea, providing students with a deeper understanding of these remarkable organisms.

Firstly, let's delve into the unique characteristics of Archaea. Unlike bacteria, Archaea possess a distinct cell membrane composed of ether lipids, which allows them to thrive in extreme environments, such as hot springs, salt pans, and deep-sea hydrothermal vents. This ability to withstand harsh conditions makes them incredibly resilient and adaptable. Additionally, Archaea lack peptidoglycan in their cell walls, setting them apart from bacteria.

Furthermore, Archaea exhibit a wide range of metabolic capabilities. Some are capable of performing photosynthesis, similar to plants, while others are chemotrophs, relying on chemical reactions to obtain energy. Methanogens, a type of Archaea, produce methane gas as a byproduct of their metabolic processes, playing a crucial role in the global carbon cycle.

Now, let's explore the classification of Archaea. Archaea are classified into three main groups: Euryarchaeota, Crenarchaeota, and Korarchaeota. Euryarchaeota includes organisms that are commonly found in extreme environments, such as high temperatures, high salinity, and acidic conditions. Crenarchaeota, on the other hand, are often found in more moderate environments and play significant roles in the nitrogen cycle. Lastly, Korarchaeota is a smaller group that has only been discovered in geothermal environments so far.

It is important to note that our understanding of Archaea is still evolving, and new discoveries continue to reshape our knowledge of these microorganisms. The study of Archaea not only expands our understanding of microbial diversity but also provides insights into the origins of life on Earth.

In conclusion, Archaea are a remarkable group of microorganisms with unique characteristics and classification. Their ability to survive in extreme environments and their diverse metabolic capabilities make them fascinating subjects of study. By exploring the distinctive features and classification of Archaea, students can gain a deeper appreciation for the complexity and diversity of the microbial world.

Eukaryotic Microorganisms: Protists and Fungi

In the vast world of microbiology, eukaryotic microorganisms play a crucial role. These organisms, known as protists and fungi, are not only fascinating but also have a significant impact on various aspects of life on Earth. In this subchapter, we will delve into the intriguing world of eukaryotic microorganisms, exploring their characteristics, roles, and importance.

Protists, often referred to as the "odds and ends" of the microbial world, are a diverse group of eukaryotic microorganisms. They can be unicellular, colonial, or multicellular, and exhibit a wide range of shapes, sizes, and modes of nutrition. Protists can be found in almost every habitat on Earth, from freshwater ponds and oceans to soil and the digestive tracts of animals. Some protists, such as algae, are photosynthetic and are vital for the production of oxygen and the cycling of nutrients in aquatic ecosystems. Others, like the parasitic protists, can cause diseases in both humans and animals.

Fungi, on the other hand, are a group of eukaryotic microorganisms that are well-known for their ability to decompose organic matter. They play a crucial role in nutrient cycling by breaking down dead organic material and recycling it back into the ecosystem. Fungi can be found in diverse habitats, including soil, water, and even inside the bodies of other organisms. They can be unicellular, such as yeasts, or multicellular, forming complex structures like mushrooms. Fungi are not only important for the environment but also have practical applications in industries such as food production, medicine, and biotechnology.

Understanding the characteristics and roles of protists and fungi is essential for students studying biology. By learning about their diversity, ecological importance, and potential applications, students can gain a deeper understanding of the invisible world of microorganisms and their impact on the planet. This knowledge can help students appreciate the intricate interconnections between organisms and ecosystems, as well as the potential benefits and challenges that arise from interactions with eukaryotic microorganisms.

In conclusion, the subchapter on eukaryotic microorganisms, specifically protists and fungi, provides students with a comprehensive overview of these fascinating organisms. From their diverse forms and habitats to their vital ecological roles and practical applications, protists and fungi have captivated scientists and researchers for centuries. By delving into their world, students can gain a deeper appreciation for the invisible yet essential organisms that shape our planet.

Chapter 5: Microbial Metabolism

Enzymes and Metabolic Pathways

In the realm of microbiology, enzymes and metabolic pathways play a crucial role in understanding the intricate workings of the invisible world. For students delving into the captivating field of biology, comprehending the significance of enzymes and metabolic pathways is essential in unraveling the mysteries of microorganisms.

Enzymes, often referred to as the catalysts of life, are remarkable molecules that facilitate and accelerate chemical reactions within cells. These biological catalysts are responsible for breaking down complex molecules into simpler ones, as well as building complex molecules from simpler ones. Without enzymes, the survival and functioning of microorganisms would be severely compromised.

Metabolic pathways, on the other hand, are a series of interconnected enzymatic reactions that occur within a cell, leading to the production of energy and the synthesis of essential biomolecules. These pathways are like intricate road maps, guiding the flow of chemical reactions and ensuring the optimal utilization of resources.

One of the most well-known metabolic pathways is glycolysis, which occurs in the cytoplasm of cells. Glycolysis is the first step in the breakdown of glucose, a fundamental sugar molecule, and it generates energy in the form of adenosine triphosphate (ATP). Understanding the details of glycolysis provides students with insights into the energy-generating processes of microorganisms.

Enzymes and metabolic pathways are highly regulated in microorganisms to maintain a delicate balance between energy production and consumption. For instance, feedback inhibition is a mechanism where the end product of a metabolic pathway inhibits the activity of the enzyme responsible for its production, ensuring that resources are not wasted.

Moreover, enzymes and metabolic pathways are not only crucial for energy production but also for the synthesis of biomolecules essential for the survival of microorganisms, such as amino acids, nucleotides, and lipids. The understanding of these pathways enables students to recognize the intricate connections between different cellular processes and appreciate the complexity of microbial life.

In conclusion, enzymes and metabolic pathways form the backbone of cellular processes in microorganisms. As students exploring the captivating world of biology, comprehending the significance of these elements is vital. By understanding the role of enzymes in catalyzing reactions and the interconnectedness of metabolic pathways, students gain a deeper appreciation for the invisible world and the remarkable mechanisms that drive microorganisms.

Energy Production: Aerobic and Anaerobic Respiration

In the world of microbiology, energy production is a vital process that allows organisms to carry out various functions necessary for their survival. Two main methods of energy production are aerobic and anaerobic respiration. This subchapter will explore these processes and their significance in the field of biology.

Aerobic respiration is a process that requires oxygen to produce energy. It is the most common method of energy production in higher organisms, including humans. During aerobic respiration, glucose is broken down into carbon dioxide and water, releasing a significant amount of energy. This energy is captured in the form of adenosine triphosphate (ATP), the universal encrgy currency of cells. ATP is used by cells to perform essential functions, such as growth, movement, and reproduction.

Anaerobic respiration, on the other hand, does not require oxygen and is commonly observed in microorganisms living in environments devoid of oxygen. This process allows these organisms to produce energy in the absence of oxygen. While it is less efficient than aerobic respiration, anaerobic respiration still provides a viable energy source for many microorganisms. The byproducts of anaerobic respiration can vary depending on the organism, with some producing ethanol or lactic acid as waste products.

Understanding the differences between aerobic and anaerobic respiration is crucial in the field of biology. It helps researchers comprehend the adaptability of various organisms to different environments. For instance, the study of anaerobic bacteria and their unique energy production processes has led to significant advancements in bioremediation, where these bacteria are employed to clean up environmental pollutants.

Moreover, the knowledge of energy production processes is essential for studying diseases caused by pathogenic microorganisms. By understanding how these organisms generate energy, scientists can identify potential targets for drug development and devise strategies to disrupt their energy production, thus inhibiting their growth and survival.

For students interested in microbiology and biology, delving into the intricacies of energy production provides a foundation for understanding the invisible world of microorganisms. It allows students to comprehend the fundamental processes that drive life on Earth, paving the way for further exploration in fields such as genetics, metabolism, and environmental science.

In conclusion, the study of energy production through aerobic and anaerobic respiration is of great importance in the field of biology. By understanding these processes, students can gain insights into the adaptability of microorganisms, advancements in bioremediation, and the development of strategies to combat pathogenic microorganisms. This knowledge forms the building blocks for further exploration in the fascinating world of microbiology.

Fermentation and its Importance in Microbial Metabolism

In the invisible world of microbiology, fermentation plays a crucial role in the metabolism of microorganisms. This subchapter aims to demystify the concept of fermentation and highlight its significance in biology.

Fermentation is a metabolic process that occurs in the absence of oxygen, known as anaerobic conditions. It involves the conversion of organic compounds into simpler substances, such as alcohol and lactic acid, by microorganisms like bacteria and yeast. This process has been utilized by humans for thousands of years in various applications, including the production of food, beverages, and biofuels.

One of the primary reasons fermentation is so important in microbial metabolism is its ability to generate energy without the presence of oxygen. In anaerobic environments, such as the deep ocean or the digestive system of animals, microorganisms rely on fermentation to break down complex molecules and release energy. This energy is essential for the survival and growth of these microorganisms in environments where oxygen is limited or absent.

Moreover, fermentation is widely used in the food industry to produce a range of products. For example, the fermentation of milk by lactic acid bacteria is responsible for the production of yogurt and cheese. The fermentation of grapes by yeast results in the production of wine. Through controlled fermentation processes, microorganisms transform raw materials into delicious and nutritious foods, enhancing their flavor, texture, and nutritional value.

Beyond food production, fermentation also holds great potential in the field of biofuels. With the increasing demand for alternative and sustainable energy sources, researchers are exploring the use of fermentation to produce biofuels such as ethanol. By fermenting plant-based materials like corn or sugarcane, microorganisms can convert the complex sugars into ethanol, which can be used as a renewable fuel source.

Understanding the importance of fermentation in microbial metabolism is crucial for students studying biology. It provides insights into the diverse metabolic capabilities of microorganisms and their applications in various fields. From the production of everyday foods to the development of sustainable energy sources, fermentation is a fundamental process that shapes the invisible world of microbiology.

In conclusion, fermentation is a vital process in microbial metabolism that occurs in anaerobic conditions. Its ability to generate energy without oxygen, along with its applications in food production and biofuel development, highlights its significance in the field of biology. By exploring the role of fermentation, students can gain a deeper understanding of the invisible world of microorganisms and their impact on our daily lives.

Photosynthesis and the Role of Microbes

Photosynthesis is one of the most fundamental and important processes in biology, as it is responsible for converting sunlight into energy that sustains life on Earth. While most of us associate photosynthesis with plants, it is important to recognize the crucial role that microbes play in this vital process.

Microbes, including bacteria and algae, are essential for photosynthesis to occur. They not only contribute to the diversity of life on our planet but also perform key functions in the process of converting solar energy into chemical energy. In fact, without these microorganisms, photosynthesis would not be as efficient or widespread as it is today.

One important group of microbes involved in photosynthesis is cyanobacteria, also known as blue-green algae. These organisms are capable of photosynthesis and were the first to evolve oxygenic photosynthesis, which produces oxygen as a byproduct. Cyanobacteria played a significant role in shaping the Earth's atmosphere, gradually increasing the oxygen levels and paving the way for the development of more complex life forms.

In addition to cyanobacteria, other microbes such as green algae and diatoms also contribute to photosynthesis. These microorganisms are found in various environments, including oceans, lakes, and even desert soils. They utilize sunlight, water, and carbon dioxide to produce glucose, which serves as an energy source for their growth and survival.

Microbes also play a crucial role in the global carbon cycle. Through photosynthesis, they remove carbon dioxide from the atmosphere, helping to mitigate the effects of climate change. Furthermore, the organic matter produced by photosynthetic microbes serves as a food source for other organisms, contributing to the overall biodiversity of ecosystems.

Understanding the role of microbes in photosynthesis is not only essential for studying basic biological processes but also has practical implications. For example, researchers are exploring ways to harness the power of photosynthetic microbes to develop sustainable and renewable energy sources, such as biofuels.

In conclusion, photosynthesis is a complex process that relies on the involvement of various microbes. These microscopic organisms not only contribute to the diversity of life on Earth but also play a vital role in converting sunlight into energy. By recognizing and studying the role of microbes in photosynthesis, students can gain a deeper understanding of the invisible world of microbiology and its significance in biology.

Chapter 6: Microbial Genetics

DNA Structure and Replication in Microorganisms

In the microscopic world of microorganisms, DNA plays a crucial role in their existence and survival. Understanding the structure and replication of DNA is essential for students in the field of biology, as it provides invaluable insights into the invisible world of microorganisms.

The structure of DNA, or deoxyribonucleic acid, is a double helix that resembles a twisted ladder. It consists of two strands made up of nucleotides, which are the building blocks of DNA. Each nucleotide comprises a sugar molecule, a phosphate group, and a nitrogenous base. The nitrogenous bases, adenine (A), thymine (T), cytosine (C), and guanine (G), pair up in a specific manner: A always pairs with T, and C always pairs with G. This base pairing forms the rungs of the DNA ladder.

Replication of DNA in microorganisms is a vital process that ensures the transmission of genetic information from one generation to the next. Before a cell divides, its DNA must be replicated to provide each new cell with a complete set of genetic instructions. This process is known as DNA replication.

During replication, the two strands of DNA separate, and each serves as a template for the synthesis of a new complementary strand. Enzymes, such as DNA polymerase, catalyze the addition of nucleotides to form the new DNA strands. The result is two identical DNA molecules, each containing one original strand and one newly synthesized strand. This semi-conservative replication ensures the preservation of genetic information and allows for genetic diversity through mutation.

Microorganisms, such as bacteria and viruses, rely on DNA replication to reproduce and adapt to changing environments. Understanding the mechanisms of DNA replication in microorganisms is crucial for studying their evolution, pathogenicity, and potential applications in various fields, including medicine and biotechnology.

By comprehending the structure and replication of DNA in microorganisms, students can gain a deeper understanding of the invisible world of microorganisms and their impact on biological systems. This knowledge can serve as a foundation for further exploration into topics such as genetic engineering, antibiotic resistance, and the role of microorganisms in human health and disease.

In conclusion, DNA structure and replication are fundamental concepts in microbiology, enabling students to unlock the mysteries of the invisible world. By delving into the intricate details of DNA, students can uncover the secrets that govern the behavior and characteristics of microorganisms, paving the way for advancements in various scientific disciplines.

Gene Expression: Transcription and Translation

In the fascinating world of microbiology, understanding the processes of gene expression is crucial to unraveling the mysteries of life itself. Gene expression refers to the process by which information encoded in the DNA molecule is converted into functional products, such as proteins or RNA molecules. This subchapter will delve into the intricate mechanisms of transcription and translation, the two key steps involved in gene expression.

Transcription, the first step in gene expression, occurs in the nucleus of eukaryotic cells and the cytoplasm of prokaryotic cells. It involves the synthesis of an RNA molecule using a DNA template. The enzyme responsible for this crucial process is called RNA polymerase, which binds to the DNA at a specific region known as the promoter. Once bound, RNA polymerase unwinds the DNA double helix and synthesizes a complementary RNA strand from the template strand of DNA. The newly synthesized RNA molecule, known as messenger RNA (mRNA), carries the genetic information from the DNA to the ribosomes, where translation takes place.

Translation, the second step in gene expression, occurs in the ribosomes, the cellular machinery responsible for protein synthesis. This process involves the conversion of the mRNA sequence into an amino acid sequence, which ultimately leads to the synthesis of a functional protein. The mRNA molecule is read in sets of three nucleotides, known as codons, each of which codes for a specific amino acid. Transfer RNA (tRNA) molecules, with their anticodon sequences complementary to the codons on the mRNA, bring the corresponding amino acids to the ribosomes. With the help of ribosomal RNA (rRNA), the ribosome ensures that the correct amino acids are linked together to form a polypeptide chain, which will ultimately fold into a functional protein.

Understanding the processes of transcription and translation is essential for comprehending the intricate workings of the cell. These processes not only play a crucial role in the expression of genes but also regulate the overall functioning of living organisms. Through gene expression, cells can respond to their environment, adapt to changing conditions, and carry out specialized functions.

As students of biology, grasping the mechanisms of gene expression will provide a solid foundation for the study of various biological processes, including development, evolution, and disease. By unraveling the mysteries of transcription and translation, we can unlock the secrets of life's invisible world and gain a deeper appreciation for the complexity and beauty of the microbial realm.

Genetic Transfer Mechanisms in Bacteria

In the fascinating world of microbiology, bacteria play a crucial role in various biological processes. One such process is genetic transfer, which allows bacteria to exchange genetic material and acquire new traits. Understanding the mechanisms behind genetic transfer is essential for students studying biology, as it provides insights into the evolution and adaptation of bacteria.

Bacteria have developed several ways to transfer genetic material, and three primary mechanisms are widely recognized: transformation, transduction, and conjugation.

Transformation is a process where bacteria naturally take up DNA fragments from their environment. These fragments can come from dead bacteria or secreted by other living bacteria. Once the DNA enters the bacterial cell, it can integrate into the recipient's genome, leading to the acquisition of new genetic traits. This mechanism has been extensively studied in bacteria like Streptococcus pneumoniae and Haemophilus influenzae, where it plays a crucial role in their pathogenicity.

Transduction, on the other hand, is a genetic transfer mechanism mediated by bacteriophages, which are viruses that infect bacteria. During the lytic cycle of a bacteriophage infection, the phage mistakenly packages bacterial DNA into its viral capsid instead of its own DNA. When this phage infects another bacterium, it injects the bacterial DNA, leading to its integration into the recipient's genome. This process allows for the transfer of specific genes between bacteria, contributing to the spread of antibiotic resistance genes, for example.

Conjugation is a direct transfer of genetic material between two bacterial cells. It involves a specialized structure called a pilus, which forms a physical bridge between the donor and recipient cells. The donor cell contains a plasmid, a small circular DNA molecule that carries additional genes. Through the pilus, the plasmid is transferred from the donor to the recipient cell, where it can replicate and express its genes. Conjugation is a crucial mechanism in the spread of antibiotic resistance genes among bacteria, making it an essential area of study for researchers and students alike.

Understanding these mechanisms of genetic transfer in bacteria opens the door to various applications in medicine, biotechnology, and evolutionary biology. The ability of bacteria to acquire new genetic traits through these mechanisms contributes to their adaptability and survival in different environments. By studying genetic transfer, students can gain insights into the mechanisms behind bacterial evolution and the potential implications for human health.

Overall, the study of genetic transfer mechanisms in bacteria is a fascinating field within microbiology. It provides students with a deeper understanding of the invisible world of microorganisms and the complex processes that shape their genetic makeup.

Genetic Engineering and its Applications in Microbiology

In recent years, the field of genetic engineering has revolutionized various scientific disciplines, including microbiology. This subchapter aims to introduce students to the fascinating world of genetic engineering and its myriad applications in the field of microbiology.

Genetic engineering involves manipulating an organism's genetic material to modify its characteristics or introduce new traits. In the context of microbiology, scientists have harnessed this powerful tool to better understand microorganisms, develop new treatments, and improve various industrial processes.

One of the primary applications of genetic engineering in microbiology is the production of recombinant proteins. By introducing genes coding for specific proteins into bacteria or yeast, scientists can use these microorganisms as miniature factories to produce valuable substances such as insulin, growth hormones, and vaccines. This technique, known as recombinant DNA technology, has revolutionized the pharmaceutical industry, enabling the mass production of life-saving medications.

Moreover, genetic engineering has opened up new possibilities for disease diagnosis and treatment. Microorganisms can be genetically modified to express fluorescent proteins, allowing scientists to track their movement and study their behavior. This knowledge has been pivotal in developing new diagnostic tools and therapies for infectious diseases caused by bacteria, viruses, and fungi.

Another exciting application of genetic engineering in microbiology is the development of genetically modified crops. By introducing desirable traits into plants, such as resistance to pests or tolerance to drought, scientists can improve crop yields and reduce the need for harmful pesticides. This technology has the potential to address global food security challenges and promote sustainable agriculture.

Furthermore, genetic engineering has facilitated the study of microbial genetics and evolution. By manipulating genes in microorganisms, scientists can identify the functions of specific genes and gain insights into the complex metabolic pathways that allow microbes to thrive in diverse environments. This knowledge is invaluable in understanding microbial evolution and developing strategies to combat antibiotic resistance.

In conclusion, genetic engineering has revolutionized the field of microbiology, offering endless possibilities for scientific exploration and practical applications. From the production of life-saving drugs to the development of genetically modified crops, this technology has the potential to address pressing global challenges. By understanding the principles and applications of genetic engineering, students can embark on a journey to unravel the mysteries of the invisible world and contribute to advancements in biology and microbiology.

Chapter 7: Microbial Interactions and Ecology

Symbiotic Relationships: Mutualism, Commensalism, and Parasitism

In the vast and diverse world of microbiology, organisms often rely on each other for survival through various types of symbiotic relationships. These relationships, which are classified as mutualism, commensalism, and parasitism, play a crucial role in shaping the invisible world and are essential to understanding the intricate dynamics of biology.

Mutualism is a symbiotic relationship where both organisms involved benefit from their association. This mutually beneficial interaction allows them to thrive together. An excellent example of mutualism is the relationship between plants and their pollinators. Bees, for instance, collect nectar from flowers for food while simultaneously spreading pollen, aiding in the plant's reproduction. Both the bees and the plants rely on each other for their survival and reproductive success.

Commensalism refers to a symbiotic relationship where one organism benefits, while the other remains unaffected. In this type of relationship, one organism utilizes the resources or shelter provided by the other without causing any harm or benefit in return. A classic example of commensalism is the relationship between cattle egrets and grazing cattle. The egrets feed on insects stirred up by the cattle while the cattle are unaffected by their presence. This relationship demonstrates how one organism can take advantage of another's activities without causing any harm.

On the other end of the spectrum, parasitism is a symbiotic relationship where one organism, the parasite, benefits at the expense of the host organism. Parasites rely on their hosts for nutrients and shelter, often causing harm or disease. A well-known example of parasitism is the relationship between ticks and mammals. Ticks feed on the blood of their hosts, which can lead to discomfort, disease transmission, and even death in severe cases. This parasitic interaction highlights the negative impact that parasites can have on their hosts.

Understanding these symbiotic relationships is crucial for students studying biology as they provide insights into the intricate connections between organisms. By studying mutualism, commensalism, and parasitism, students gain a deeper understanding of how organisms depend on each other for survival. These relationships not only shape the invisible world of microorganisms but also play a significant role in the ecosystems and the overall health of our planet.

In conclusion, symbiotic relationships are fundamental to microbiology, biology, and the natural world as a whole. Mutualism, commensalism, and parasitism demonstrate the dynamic interactions between organisms, ranging from mutually beneficial to harmful associations. By exploring these symbiotic relationships, students can unravel the complexities of the invisible world and gain a greater appreciation for the interconnectedness of all living organisms.

Microbial Interactions in the Environment

Microbiology Demystified: A Student's Guide to Understanding the Invisible World

Welcome to the fascinating world of microbial interactions in the environment! In this subchapter, we will explore the intricate relationships that microorganisms have with each other and their surroundings. Understanding these interactions is crucial in the field of biology, as it sheds light on the roles microorganisms play in maintaining the delicate balance of ecosystems.

Microbial interactions occur at various levels, from individual cells to entire communities. One of the most common forms of interaction is mutualism, where different microbial species cooperate and benefit from each other's presence. For example, nitrogen-fixing bacteria form a symbiotic relationship with legume plants, providing them with essential nitrogen compounds while receiving a source of energy in return. This type of interaction is vital for nutrient cycling in the environment.

On the other hand, some microbial interactions involve competition for limited resources. Microbes are constantly at odds with each other, fighting for space, nutrients, and other essentials. This competition can have profound effects on the dynamics of microbial communities. The ability of certain microbial species to outcompete others can influence the overall composition and stability of an ecosystem.

Microbial interactions can also lead to predation and parasitism. Certain microorganisms have evolved mechanisms to prey upon or infect other microbes, effectively utilizing them as a source of nutrients. These interactions help regulate population sizes and prevent the dominance of a single species. Moreover, they can provide valuable insights into the development of antimicrobial agents and the control of infectious diseases.

In addition to interactions between different microbial species, microorganisms also interact with their physical environment. They can alter the chemistry of their surroundings, release compounds that modify the growth of other organisms, and even form biofilms, which are complex communities of microorganisms attached to surfaces. Biofilms can have profound impacts on human health, as they can colonize medical devices, leading to infections that are difficult to treat.

Understanding microbial interactions in the environment is crucial not only for biology students but also for researchers and professionals in various fields. By studying these interactions, we can gain insights into the functioning of ecosystems, the development of diseases, and the potential applications of microorganisms in biotechnology. So, let's embark on this exciting journey to unravel the invisible world of microbial interactions and appreciate the incredible impact these tiny organisms have on our lives!

Microbes and Human Health: Normal Flora and Pathogens

In the vast world of microbiology, the intricate relationship between microbes and human health is a captivating topic. This subchapter aims to shed light on the role of microbes in our bodies, specifically focusing on the concept of normal flora and pathogens. Understanding this dynamic is crucial for students delving into the field of biology.

Firstly, it is important to grasp the concept of normal flora. Our bodies are home to trillions of microbes, collectively known as the human microbiota. These microbes reside on our skin, in our mouths, and within our gastrointestinal and genitourinary tracts, among other areas. Contrary to what one might expect, the majority of these microbes are not harmful. In fact, they play a vital role in maintaining our overall health. They assist in digestion, produce essential vitamins, and help regulate our immune system. This symbiotic relationship between humans and microbes is a fascinating area of study in microbiology.

However, not all microbes are friendly. Some have the potential to cause harm and are referred to as pathogens. Pathogens are disease-causing microorganisms that can disrupt the delicate equilibrium of our normal flora. These include bacteria, viruses, fungi, and parasites. Pathogens can cause a range of illnesses, from mild infections to severe diseases. Understanding the mechanisms by which these microbes invade our bodies and cause disease is crucial for developing effective treatments and preventive measures.

In this subchapter, students will delve into the fascinating world of pathogens, learning about their modes of transmission, virulence factors, and the ways in which our immune system responds to their presence. The subchapter will also explore how our normal flora can help protect against pathogenic invasion, acting as a barrier against potential harm.

Moreover, the subchapter will highlight the importance of maintaining a healthy balance of normal flora. Factors such as antibiotic use, a poor diet, and stress can disrupt this delicate equilibrium, leading to an overgrowth of harmful microbes. Students will learn about the consequences of such dysbiosis and the potential health implications it may have.

Ultimately, this subchapter aims to provide students with a solid foundation in understanding the intricate relationship between microbes and human health. By exploring the concepts of normal flora and pathogens, students will gain valuable insights into the invisible world of microbiology and its impact on our everyday lives.

Biogeochemical Cycles: Microbes' Role in Nutrient Cycling

Microbes are the unsung heroes of the invisible world, playing a vital role in the biogeochemical cycles that sustain life on our planet. These microscopic organisms, such as bacteria, fungi, and archaea, are responsible for cycling nutrients through various ecosystems, ensuring the availability of essential elements for all living organisms. In this subchapter, we will explore the fascinating world of biogeochemical cycles and delve into the crucial role that microbes play in nutrient cycling.

Nutrient cycling is the process by which essential elements, such as carbon, nitrogen, phosphorus, and sulfur, are continuously recycled through different compartments of the environment. These elements are essential for the growth and survival of all living organisms, including plants, animals, and even microbes themselves. Without efficient nutrient cycling, ecosystems would cease to function, and life as we know it would not be possible.

Microbes are at the forefront of nutrient cycling, driving the transformation and recycling of organic and inorganic matter. For instance, in the carbon cycle, microbes break down complex organic compounds, such as dead plant and animal matter, into simpler forms through a process called decomposition. This decomposition releases carbon dioxide back into the atmosphere, which can then be utilized by plants during photosynthesis to produce more organic matter. The carbon cycle is a crucial component in regulating Earth's climate and maintaining the balance of greenhouse gases.

Similarly, microbes play a pivotal role in the nitrogen cycle. Nitrogen is an essential nutrient for the production of proteins and DNA in all organisms. However, most organisms cannot directly utilize atmospheric nitrogen, which is abundant but chemically inert. Microbes, known as nitrogen-fixing bacteria, have the unique ability to convert atmospheric nitrogen into a usable form through a process called nitrogen fixation. This fixed nitrogen can then be taken up by plants and other organisms, ultimately returning to the soil through decomposition.

Phosphorus and sulfur cycles also rely on microbial activity. Microbes are involved in the breakdown of organic compounds containing phosphorus and sulfur, releasing these elements back into the environment. They also play a vital role in the transformation of sulfur compounds, such as sulfate reduction, which is essential for the cycling of sulfur between the atmosphere, water bodies, and land.

Understanding the role of microbes in nutrient cycling is crucial for students studying biology. It highlights the intricate interconnectedness of all living organisms and their environment. By studying the processes and interactions within biogeochemical cycles, students can gain a deeper appreciation for the delicate balance that sustains life on Earth.

In conclusion, microbes are the unsung heroes of biogeochemical cycles, playing a vital role in nutrient cycling. Their ability to transform and recycle essential elements ensures the availability of nutrients for all organisms. By studying the role of microbes in these cycles, students can gain a better understanding of the invisible world and its impact on the biology of our planet.

Chapter 8: Immunology and Host-Microbe Interactions

The Immune System: Innate and Adaptive Immunity

In the intricate world of microbiology, understanding the immune system is crucial. Our immune system serves as our body's defense against harmful pathogens, viruses, and bacteria that continuously try to invade our bodies. This subchapter will shed light on the immune system's two primary lines of defense: innate and adaptive immunity. By delving into the depths of these immune responses, students studying biology will gain a comprehensive understanding of how our bodies protect themselves from the invisible world of microorganisms.

The immune system's first line of defense is the innate immunity. This is the body's rapid response that provides immediate protection against a wide range of pathogens. Innate immunity is not specific to any particular pathogen and includes physical barriers like the skin, mucous membranes, and cilia, which trap and expel foreign invaders. Additionally, chemical barriers such as stomach acid and antimicrobial proteins in bodily secretions help neutralize pathogens. Innate immunity also involves specialized cells like macrophages, neutrophils, and natural killer cells that engulf and destroy foreign substances.

The second line of defense is known as adaptive immunity, which is a more specific and targeted response. Adaptive immunity recognizes and remembers specific pathogens, allowing the body to mount a stronger defense upon subsequent encounters. This immunity is acquired during an individual's lifetime through exposure to pathogens or vaccination. Adaptive immunity involves the production of antibodies by B cells, which bind to and neutralize specific pathogens. T cells, another crucial component of adaptive immunity, directly attack infected cells. This response is highly sophisticated and can generate immunological memory, ensuring a faster and more effective response to future infections.

Understanding the interplay between innate and adaptive immunity is essential for comprehending the body's intricate defense mechanisms against pathogens. The immune system's ability to differentiate between self and non-self is a remarkable feat, and studying this phenomenon can help students appreciate the complexity and efficiency of our bodies.

In conclusion, the immune system is a fascinating topic in the field of microbiology. This subchapter on innate and adaptive immunity provides students with a comprehensive overview of the body's defense mechanisms against pathogens. By understanding how the immune system works, students studying biology can gain a deeper appreciation for the invisible world of microorganisms and the remarkable ways our bodies protect us.

Host-Microbe Interactions: Infection and Disease

In the captivating world of microbiology, one of the most fascinating areas of study is the intricate relationship between hosts and microbes. This subchapter, "Host-Microbe Interactions: Infection and Disease," will delve into the captivating dynamics of this relationship, shedding light on the causes and consequences of infections and diseases.

Microbes, although they may be invisible to the naked eye, play a significant role in the lives of all living organisms, including humans. They can exist within or on the body, forming complex communities known as microbiota. This community of microbes, collectively referred to as the human microbiome, has a profound impact on our health and well-being.

However, not all microbe-host interactions are beneficial. Some microbes have evolved mechanisms to invade and colonize their hosts, leading to infections and diseases. This subchapter will explore the intricacies of how these interactions occur and the consequences they have on the host.

To understand infections, we will delve into the concept of pathogenicity, which refers to the ability of a microbe to cause disease. We will explore how microbes breach the barriers of the host's immune system, either by evading or overpowering it. We will also discuss the various modes of transmission that allow microbes to spread from one host to another, emphasizing the importance of hygiene and preventive measures.

Furthermore, this subchapter will provide insights into the different types of diseases caused by microbes. We will explore the mechanisms by which microbes damage host tissues and disrupt normal physiological processes. Topics like virulence factors, toxins, and immune responses will be explored to help students comprehend the complexity of infection and disease.

To create a comprehensive understanding of host-microbe interactions, this subchapter will also touch upon the concept of antibiotics and antimicrobial resistance. We will discuss how overuse and misuse of antibiotics have led to the emergence of drug-resistant microbes, posing a significant challenge to public health.

By the end of this subchapter, students will have a deeper understanding of the fascinating world of host-microbe interactions, infection, and disease. They will grasp the importance of maintaining a healthy microbiome, practicing good hygiene, and adopting responsible antibiotic use. This knowledge will equip them to become informed individuals in the field of biology, ready to tackle the challenges posed by infectious diseases and contribute to the development of innovative solutions to combat them.

Vaccines and the Importance of Immunizations

In the vast field of biology, one area that has undoubtedly revolutionized the way we prevent and combat infectious diseases is the development of vaccines. Vaccines are a crucial tool in protecting ourselves and our communities from a range of harmful pathogens. By understanding the importance of immunizations, students can grasp the significant role vaccines play in maintaining public health.

Vaccines work by stimulating the immune system to recognize and remember specific pathogens. They contain weakened or inactivated forms of the disease-causing microorganisms or their parts, such as proteins or sugars. When a vaccine is administered, the immune system mounts a defense against these harmless components, effectively priming itself for future encounters with the real pathogen. This prepares the immune system to respond rapidly and effectively, preventing the development of severe illness or complications.

Immunizations have had a profound impact on society. Diseases like smallpox, which once plagued humanity, have been eradicated through widespread vaccination campaigns. Polio, measles, and rubella are other examples of diseases that have been significantly controlled or eliminated in many parts of the world due to vaccination efforts. By getting vaccinated, students not only protect themselves but also contribute to the collective immunity of their community, preventing the spread of diseases to vulnerable individuals who may be unable to receive vaccines due to age, health conditions, or other reasons.

Despite the undeniable success of vaccines, skepticism and misinformation surrounding immunizations persist. It is important for students to critically evaluate sources of information and rely on reputable scientific research. Vaccines undergo rigorous testing and monitoring before they are approved for use, ensuring their safety and efficacy. The scientific consensus overwhelmingly supports the use of vaccines as a highly effective means of disease prevention.

Moreover, vaccines have saved countless lives and prevented untold suffering throughout history. They are a testament to the power of scientific advancements and the dedication of researchers and healthcare professionals. By understanding the science behind vaccines and their role in public health, students can make informed decisions for themselves and advocate for evidence-based policies that promote immunization.

In conclusion, vaccines and the importance of immunizations are vital topics for students studying biology. By learning about vaccines, students can appreciate the impact they have had on controlling infectious diseases and preventing outbreaks. Moreover, they can understand the significance of individual and community responsibility in ensuring the success of vaccination programs. Armed with knowledge, students can become advocates for immunizations, contributing to a healthier and safer world for all.

Microbial Evasion Mechanisms and Antimicrobial Resistance

In the exciting world of microbiology, understanding the mechanisms by which microbes evade our immune system and develop resistance to antimicrobial agents is of utmost importance. This subchapter will delve into the fascinating realm of microbial evasion and the alarming rise of antimicrobial resistance, providing students with a comprehensive understanding of these critical topics.

Microbes, such as bacteria and viruses, have evolved numerous strategies to evade our immune system's defenses. These evasion mechanisms enable them to establish infections, persist in the host, and cause diseases. One such mechanism is antigenic variation, where microbes change the surface proteins frequently, making it challenging for our immune system to recognize and neutralize them. Another strategy is the production of virulence factors that manipulate our immune response, allowing the microbe to thrive and cause damage.

Furthermore, microbes can also form biofilms, which are communities of cells encapsulated in a protective matrix. Biofilms provide a physical barrier against antimicrobial agents and make it difficult for our immune cells to reach and eliminate the microbes. Through these evasion mechanisms, microbes can establish chronic infections and become resistant to our immune system's attacks.

The alarming rise of antimicrobial resistance poses a significant threat to public health. Overuse and misuse of antibiotics have accelerated the development of resistant strains of bacteria, rendering many drugs ineffective. This subchapter will explore the mechanisms by which bacteria acquire resistance genes, such as through mutations or horizontal gene transfer. Students will learn about the importance of responsible antibiotic use to slow down the emergence of resistance and preserve the effectiveness of these life-saving drugs.

Moreover, the subchapter will discuss the challenges in developing new antimicrobial agents to combat resistant microbes. Students will gain insights into alternative approaches, including the use of phage therapy, which utilizes viruses that specifically target and kill bacteria.

Understanding microbial evasion mechanisms and antimicrobial resistance is crucial for students studying biology. It equips them with the knowledge needed to tackle the challenges posed by infectious diseases and develop strategies to prevent and control them effectively. By exploring these topics, students will gain a deeper appreciation for the complex and ever-evolving world of microbiology.

Chapter 9: Applied Microbiology

Food Microbiology: Fermented Foods and Food Safety

Fermented foods have been consumed by humans for centuries and are an integral part of various cultures around the world. From sauerkraut to yogurt, these foods undergo a natural process of fermentation that not only enhances their flavor but also improves their preservation and safety.

Fermentation is a metabolic process that converts carbohydrates, such as sugars and starches, into simpler compounds using microorganisms like bacteria, yeasts, or molds. These microorganisms play a crucial role in breaking down complex molecules and producing byproducts that give fermented foods their unique taste and texture.

One of the key benefits of fermentation is the preservation of food. The growth of harmful bacteria, like Salmonella or E. coli, is hindered by the acidic environment created during fermentation. This acidic environment is a result of organic acids produced by the microorganisms during the fermentation process. As a result, the shelf life of fermented foods is extended, reducing the risk of foodborne illnesses.

Moreover, fermented foods can also improve food safety by promoting the growth of beneficial bacteria. Probiotics, which are live microorganisms that confer health benefits when consumed in adequate amounts, are often found in fermented foods. These probiotics can help maintain a healthy gut microbiota and enhance digestion and nutrient absorption.

However, it is important to note that not all fermented foods are created equal in terms of safety. While most fermented foods are safe to consume, improper fermentation conditions or contamination can lead to foodborne illnesses. Therefore, it is essential to understand the principles of food safety and proper fermentation techniques.

To ensure the safety of fermented foods, it is crucial to maintain proper hygiene during all stages of the fermentation process. This includes using clean utensils and equipment, proper washing and handling of raw ingredients, and maintaining appropriate temperatures and fermentation times.

Additionally, regular monitoring and testing of fermented foods for microbial contamination is essential. This can be done through laboratory analysis or sensory evaluation to detect any signs of spoilage or pathogenic microorganisms.

In conclusion, fermented foods are not only delicious but also offer numerous benefits to our health. They are a result of the intricate relationship between microorganisms and food, showcasing the wonders of food microbiology. However, it is important for students studying biology to understand the importance of food safety and proper fermentation techniques to ensure the quality and safety of these beloved foods.

Industrial Microbiology: Biotechnology and Biofuels

In the rapidly advancing field of biotechnology, industrial microbiology plays a crucial role in harnessing the power of microorganisms for various applications. One of the most significant areas of industrial microbiology is the production of biofuels, which are sustainable alternatives to fossil fuels. This subchapter will delve into the fascinating world of industrial microbiology, with a specific focus on biotechnology and biofuels, aimed at students with a keen interest in biology.

Biotechnology, the application of biological organisms or processes for industrial purposes, relies heavily on the use of microorganisms. These small but mighty creatures possess extraordinary capabilities that can be harnessed to produce a wide range of products, including biofuels. Biofuels are derived from renewable biological resources such as plants, algae, and microorganisms and offer a sustainable and eco-friendly alternative to traditional fossil fuels.

The production of biofuels involves the microbial conversion of organic matter into usable energy sources. Microorganisms such as bacteria and yeast possess the ability to break down complex organic compounds and convert them into simpler molecules like ethanol, biodiesel, and methane. This process, known as fermentation, is facilitated by enzymes produced by these microorganisms.

In recent years, there has been a growing interest in the use of microalgae for biofuel production. These tiny aquatic plants have the ability to convert sunlight and carbon dioxide into lipids or oils that can be further processed into biodiesel. Microbes such as bacteria and fungi are also being explored for their potential in producing biofuels from various feedstocks, including agricultural waste and lignocellulosic biomass.

Apart from biofuels, industrial microbiology is involved in the production of a range of other valuable products. These include pharmaceuticals, enzymes, organic acids, vitamins, and bioplastics, among others. The field also encompasses wastewater treatment, bioremediation, and the development of microbial pesticides.

Industrial microbiology is a rapidly evolving field with immense potential for addressing the pressing challenges of sustainable energy production and environmental conservation. As students interested in biology, exploring the world of industrial microbiology can provide valuable insights into the ways in which microorganisms can be harnessed for the benefit of society and the planet.

By understanding the principles and applications of industrial microbiology, students can gain a deeper appreciation for the invisible world of microorganisms and their significant impact on various industries. Moreover, this knowledge can inspire future scientists to explore innovative solutions for a more sustainable and bio-based economy.

In the following chapters, we will delve deeper into the fascinating world of industrial microbiology, exploring the processes involved in biofuel production, the challenges faced, and the potential future developments in this exciting field.

Environmental Microbiology: Bioremediation and Waste Treatment

In the field of environmental microbiology, bioremediation and waste treatment play a crucial role in preserving and restoring our planet's health. This subchapter aims to introduce students to these fascinating concepts, highlighting the role of microorganisms in cleaning up pollutants and treating various forms of waste.

Bioremediation refers to the use of microorganisms to degrade or transform harmful substances in the environment. When human activities, such as industrial processes or accidental spills, contaminate soil, water, or air with toxic compounds, bioremediation offers a natural and sustainable solution. Microbes possess unique metabolic capabilities that allow them to break down complex pollutants into harmless byproducts. For instance, bacteria like Pseudomonas and Bacillus species can degrade petroleum hydrocarbons, including oil spills, by utilizing them as a carbon and energy source.

Waste treatment is another critical aspect of environmental microbiology. As our population grows, so does the generation of different types of waste, including organic, industrial, and municipal waste. Microorganisms are harnessed to transform these wastes into less harmful or more manageable forms. In wastewater treatment plants, microbes help break down organic matter and remove pollutants, ensuring that the water released back into the environment is safe. Similarly, in composting facilities, microorganisms decompose organic waste, producing nutrient-rich compost for agricultural use.

Understanding the principles of bioremediation and waste treatment is vital for students in the field of biology. By applying the knowledge gained, they can contribute to developing sustainable and efficient solutions for environmental problems. Moreover, these concepts align with the growing emphasis on adopting environmentally friendly practices in various industries.

As students delve deeper into this subchapter, they will explore different bioremediation strategies, such as bioaugmentation and biostimulation, and learn about the factors that influence microbial activity in waste treatment systems. They will also examine case studies showcasing successful bioremediation projects and waste treatment technologies, providing real-world examples of how microorganisms can make a positive impact on our environment.

In conclusion, environmental microbiology, specifically bioremediation and waste treatment, offers students a unique perspective on the invisible world of microorganisms. By studying the remarkable abilities of these microscopic organisms, students can gain insights into how they can contribute to a cleaner and healthier planet.

Medical Microbiology: Diagnosis and Treatment of Infectious Diseases

In the field of biology, medical microbiology holds a crucial role in understanding and combating infectious diseases. This subchapter will delve into the fascinating world of medical microbiology, exploring the methods used for diagnosing and treating these diseases that impact human health.

Diagnosis is the first step towards effective treatment. Medical microbiologists employ a variety of techniques to identify the causative agents of infectious diseases. One commonly used method is microscopy, where samples are examined under a high-powered microscope to detect the presence of bacteria, fungi, or parasites. The advent of advanced techniques, such as polymerase chain reaction (PCR) and next-generation sequencing, has revolutionized the field by providing rapid and accurate identification of microorganisms.

Once the infectious agent has been identified, the next challenge lies in treating the disease. Antibiotics, antifungals, and antiviral drugs play a vital role in combating these infections. However, the emergence of drug-resistant strains has posed a significant challenge. Students will learn about the principles of antimicrobial susceptibility testing, which helps determine the most effective drug for a particular pathogen. This knowledge is crucial in preventing the misuse of antibiotics and the development of further resistance.

Understanding the host-pathogen interaction is another critical aspect of medical microbiology. The immune response plays a pivotal role in combating infectious diseases. Students will explore the intricate mechanisms by which the immune system recognizes and eliminates invading microorganisms. They will also learn about immunizations and how they can prevent the onset of infectious diseases.

This subchapter will also cover specific infectious diseases, such as tuberculosis, malaria, HIV/AIDS, and influenza. Students will gain insights into the epidemiology, clinical manifestations, and treatment options for these diseases. Additionally, emerging infectious diseases, such as COVID-19, will be discussed to provide students with a comprehensive understanding of the challenges faced by medical microbiologists in real-time.

By the end of this subchapter, students will have a firm grasp of the principles and practices of medical microbiology. They will appreciate the importance of accurate diagnosis, the challenges posed by drug resistance, and the significance of immunizations in preventing infectious diseases. This knowledge will serve as a solid foundation for those aspiring to pursue a career in healthcare, research, or public health, where understanding and combating infectious diseases is of utmost importance.

Chapter 10: Emerging Topics in Microbiology

Microbiome and its Impact on Human Health

The human body is a complex ecosystem that houses trillions of microorganisms, collectively known as the microbiome. These microorganisms, including bacteria, fungi, viruses, and archaea, reside in various parts of our bodies, such as the skin, mouth, gut, and reproductive system. While some of these microorganisms can cause diseases, the majority of them are harmless or even beneficial to our health.

The study of the microbiome and its impact on human health has gained significant attention in recent years. Researchers have discovered that the microbiome plays a crucial role in maintaining our overall well-being. It influences our immune system, digestion, metabolism, and even mental health.

One of the most well-studied areas of the microbiome is the gut microbiota. The gut houses a diverse community of microorganisms that aid in the digestion and absorption of nutrients, produce vitamins, and help fight off harmful pathogens. Imbalances or disruptions in the gut microbiota, known as dysbiosis, have been linked to various health conditions, including obesity, diabetes, inflammatory bowel disease, and even mental disorders like depression and anxiety.

Understanding the microbiome and its impact on human health has opened up new avenues for disease prevention and treatment. Probiotics, for example, are live microorganisms that, when consumed in adequate amounts, provide health benefits to the host. These beneficial bacteria can be found in certain fermented foods like yogurt and sauerkraut, or they can be taken as supplements. By restoring the balance of the gut microbiota, probiotics have shown promising results in alleviating symptoms of gastrointestinal disorders and strengthening the immune system.

Furthermore, the study of the microbiome has also revealed the importance of a healthy lifestyle in maintaining a diverse and balanced microbial community. Factors such as diet, stress, antibiotics, and hygiene practices can significantly impact the composition of the microbiome. A diet rich in fiber and plant-based foods, for instance, promotes the growth of beneficial bacteria, while excessive use of antibiotics can disrupt the delicate balance of the microbiota.

In conclusion, the microbiome is a fascinating and intricate part of the human body that has a profound impact on our health. By understanding its role and how it can be influenced, students studying biology can gain valuable insights into the intricate workings of the invisible world. As research in this field continues to advance, it is becoming increasingly clear that nurturing a healthy microbiome is crucial for our overall well-being.

Microbial Ecology and Climate Change

Climate change is one of the most pressing global challenges of our time. It is a phenomenon that affects various aspects of our planet, including the delicate balance of microbial communities. In this subchapter, we will explore the intricate relationship between microbial ecology and climate change, shedding light on how these tiny organisms play a vital role in the Earth's climate system.

Microbes, which include bacteria, archaea, fungi, and viruses, are found in every corner of our planet. They inhabit diverse environments, from the depths of the oceans to the highest mountains, and even within our own bodies. These microscopic organisms are incredibly versatile and can adapt to changing environmental conditions, making them crucial players in climate change dynamics.

One significant aspect of microbial ecology and climate change is the role of microorganisms in the carbon cycle. Microbes are involved in both carbon storage and release processes. They decompose organic matter, releasing carbon dioxide into the atmosphere. On the other hand, certain microbial communities, such as photosynthetic bacteria and algae, absorb carbon dioxide through photosynthesis, mitigating the greenhouse effect. Understanding these processes is crucial for predicting and managing climate change.

Additionally, microbial communities are sensitive to changes in temperature and precipitation patterns caused by climate change. Alterations in these environmental factors can lead to shifts in microbial composition and diversity. Such changes can have cascading effects on ecosystems, including nutrient cycling, soil fertility, and plant health. For instance, as temperatures rise, certain pathogenic microorganisms may thrive, posing risks to human and animal health.

Moreover, microbial ecology and climate change are closely intertwined in the context of extreme weather events. Microbes play a crucial role in the breakdown of organic matter during floods, hurricanes, and other natural disasters. These events can introduce large quantities of organic material into aquatic ecosystems, which can lead to changes in microbial populations, water quality, and overall ecosystem function.

In conclusion, microbial ecology is deeply intertwined with climate change dynamics. Microbes influence the carbon cycle, respond to changing environmental conditions, and play essential roles in nutrient cycling and ecosystem functioning. Understanding the intricate relationship between microbial communities and climate change is crucial for developing strategies to mitigate and adapt to the challenges that lie ahead. As students of biology, it is essential to recognize the significant role that microbial ecology plays in our planet's climate system and to continue exploring this fascinating field to address the challenges of tomorrow.

Microbes in Space: Astrobiology and Exobiology

In the vast expanse of outer space, where the stars twinkle and planets orbit, lies a fascinating field of study known as astrobiology. This subchapter explores the intriguing world of microbes in space and the captivating science of exobiology. Aimed at students with an interest in biology, the following pages will shed light on the invisible world of microorganisms beyond our planet.

Astrobiology is the study of life and its potential existence beyond Earth. Scientists in this field investigate the conditions necessary for life to thrive in the vastness of space. One crucial aspect of astrobiology is understanding the role of microbes, tiny organisms that are invisible to the naked eye. These microorganisms, such as bacteria and fungi, may hold the key to understanding the origins of life and the possibility of life existing elsewhere in the universe.

The study of exobiology focuses on the search for extraterrestrial life forms, including microbial life. It involves analyzing samples from space missions, such as the Mars rovers, to detect any signs of microbial activity. These missions provide valuable insights into the potential habitability of other planets and moons within our solar system.

Microbes in space face extreme conditions, including harsh temperatures, high levels of radiation, and the vacuum of space. Despite these challenges, scientists have discovered that certain microorganisms are remarkably resilient and can survive in these hostile environments. Studying these extremophiles not only helps us understand the limits of life on Earth but also provides clues about the potential habitability of other celestial bodies.

Furthermore, the discovery of water on Mars and the presence of organic molecules on Saturn's moon, Enceladus, have fueled the excitement surrounding the search for extraterrestrial life. Studying the microbial ecology of extreme environments on Earth, such as deep-sea hydrothermal vents and Antarctic glaciers, can provide insights into potential habitats for life beyond our planet.

In conclusion, the field of astrobiology and exobiology offers an intriguing view into the possibility of life beyond Earth. Microbes play a crucial role in this exploration, as they provide insights into the origins of life and the potential for habitability in space. This subchapter aims to inspire students with an interest in biology to delve into the fascinating world of astrobiology and exobiology, where the invisible world of microorganisms holds the key to unraveling the mysteries of our universe.

Future Perspectives in Microbiology

As students studying the fascinating world of biology, it is crucial to understand the future perspectives in microbiology. The field of microbiology has undergone significant advancements in recent years, and it continues to hold immense potential for further exploration and discoveries. In this subchapter, we will delve into the exciting possibilities that lie ahead in the realm of microbiology.

One of the most promising areas in microbiology is the study of the human microbiome. The human body is home to trillions of microorganisms, collectively known as the microbiome, which play a crucial role in our health and well-being. Researchers are constantly uncovering the intricate relationship between the microbiomc and various diseases, paving the way for personalized medicine and targeted therapies. Understanding the human microbiome holds the key to managing and treating conditions such as obesity, autoimmune diseases, and mental health disorders.

Another exciting prospect in microbiology is the development of novel antimicrobial agents. With the rise of antibiotic-resistant bacteria, there is an urgent need for alternative treatments. Scientists are exploring new avenues, such as phage therapy, which utilizes viruses to target and eliminate specific bacteria, offering a potential solution to combat antibiotic resistance. Furthermore, the discovery of new antimicrobial compounds from natural sources, such as marine organisms and plants, holds promise for developing effective and sustainable treatments.

The field of biotechnology is also set to revolutionize various industries. Microbes have been harnessed to produce valuable compounds, including pharmaceuticals, biofuels, and enzymes. The ability to manipulate and engineer microorganisms opens up a world of possibilities, from creating environmentally friendly processes to generating renewable resources. As students interested in biology, you have the opportunity to contribute to these advancements by exploring the field of microbial biotechnology.

Additionally, the application of microbiology in environmental conservation is gaining momentum. Microbes play a vital role in maintaining ecosystem balance, and understanding their interactions can help us protect and restore damaged environments. Microbes have been used to clean up oil spills, degrade pollutants, and enhance soil fertility. Advancements in microbial ecology and bioremediation techniques will continue to shape our understanding of the environment and our ability to mitigate environmental challenges.

In conclusion, the future of microbiology is brimming with potential. From the study of the human microbiome to the development of novel antimicrobial agents, the field offers numerous avenues for exploration and innovation. As students with a passion for biology, you have the opportunity to contribute to these exciting developments and shape the future of microbiology. Embrace the invisible world of microorganisms and unlock the mysteries that lie within!

Chapter 11: Microbiology Laboratory Techniques

Aseptic Techniques and Culturing Microorganisms

In the fascinating world of microbiology, aseptic techniques and culturing microorganisms form the foundation of scientific research in the field. Whether you are a biology student just starting your journey into the invisible world or a seasoned researcher, understanding these techniques is crucial for successful experimentation and analysis.

Aseptic techniques refer to a set of practices that aim to prevent contamination when working with microorganisms. As the name suggests, it involves creating a sterile environment to ensure that only the desired microorganisms are present in the culture. This is achieved by sterilizing all equipment and surfaces, using techniques such as flaming, autoclaving, or chemical disinfection. Sterilization ensures that unwanted bacteria or fungi do not interfere with the growth of the target microorganisms.

Culturing microorganisms involves providing the ideal conditions for their growth and reproduction in a laboratory setting. This is done by creating a culture medium that contains all the necessary nutrients for the specific microorganism being studied. Different types of media, such as agar plates or liquid broths, are used depending on the requirements of the microorganism. Culturing microorganisms allows scientists to study their characteristics, behavior, and interactions.

To begin culturing microorganisms, aseptic techniques must be followed. This includes using a Bunsen burner to sterilize inoculating loops or needles before transferring the microorganisms onto the culture medium. It is crucial to maintain a sterile environment throughout the process to prevent contamination and obtain accurate results.

Once the microorganisms are inoculated onto the culture medium, they are incubated at specific temperatures and conditions that favor their growth. Incubation periods can vary depending on the microorganism, ranging from a few hours to several days. During this time, it is important to monitor the cultures for signs of growth and record any observations.

Culturing microorganisms provides invaluable insights into their biology, including their morphology, metabolic processes, and response to various environmental factors. It also allows for the identification and study of pathogenic microorganisms, aiding in the development of diagnostic tests and treatments for infectious diseases.

As a student of biology, mastering aseptic techniques and culturing microorganisms is essential. These techniques will not only be used in laboratory settings but also in various research fields such as medicine, environmental science, and biotechnology. Understanding the principles behind aseptic techniques and culturing microorganisms equips you with the skills necessary to contribute to groundbreaking discoveries and advancements in the field of microbiology.

Microscopic Examination of Microbes

In the captivating world of microbiology, understanding the invisible realm of microbes is crucial. This subchapter will delve into the fascinating techniques used to observe and study these tiny organisms under a microscope. By learning about the microscopic examination of microbes, students can unlock a deeper understanding of the invisible world that surrounds us.

Microscopy is an indispensable tool in microbiology, allowing scientists to visualize and analyze microorganisms that are otherwise invisible to the naked eye. The most commonly used microscope in microbiology is the compound light microscope. This microscope employs a series of lenses to magnify the specimen, enabling students to observe minute details of microbial structures, such as cell morphology and arrangement.

To prepare a sample for microscopic examination, a technique called a smear is typically employed. This involves spreading a thin layer of the microbial sample onto a microscope slide, which is then treated with a stain to enhance visibility. Staining techniques, such as the Gram stain, allow students to differentiate between different types of microbes based on their cell wall composition. The Gram stain, for instance, can distinguish between Gram-positive and Gram-negative bacteria, providing valuable information about their structure and behavior.

Moreover, microscopy techniques extend beyond simple staining. Phase-contrast microscopy and dark-field microscopy are two advanced techniques that enable the visualization of live, unstained microbes. Phase-contrast microscopy enhances the contrast of transparent specimens, while dark-field microscopy illuminates the specimen from the side, creating a bright image against a dark background. These techniques are particularly useful for observing delicate and fast-moving microorganisms.

In addition to traditional light microscopy, electron microscopy provides even higher magnification and resolution. Scanning electron microscopy (SEM) and transmission electron microscopy (TEM) utilize beams of electrons to generate detailed images of microbes. SEM allows for the three-dimensional visualization of the specimen's surface, while TEM provides a cross-sectional view, revealing internal structures.

By utilizing various microscopic techniques, students can explore the intricate world of microbes. The ability to observe and study these tiny organisms allows for a better understanding of their behavior, interactions, and impact on our environment. Through microscopic examination, students can pave the way for groundbreaking discoveries and advancements in the field of microbiology.

In conclusion, the microscopic examination of microbes is an essential aspect of microbiology, enabling students to explore the invisible world that surrounds us. By employing different microscopy techniques, such as compound light microscopy, electron microscopy, phase-contrast microscopy, and dark-field microscopy, students can gain insights into the structure and behavior of microorganisms. This knowledge is invaluable in fields such as biology, where understanding the intricacies of the invisible world is crucial for scientific advancements. So, grab your microscope and embark on a journey into the captivating realm of microbiology!

Staining Techniques: Gram Stain and Acid-Fast Stain

In the microscopic world of microbiology, staining techniques play a crucial role in identifying and classifying various microorganisms. Two commonly used staining methods are the Gram stain and the Acid-Fast stain. These techniques allow scientists and researchers to distinguish between different types of bacteria, aiding in the diagnosis and treatment of infectious diseases. In this subchapter, we will explore the principles and applications of these staining techniques, providing students with a comprehensive understanding of their significance in the field of biology.

The Gram stain, developed by Danish scientist Hans Christian Gram in 1884, is one of the most widely used staining techniques in microbiology. It differentiates bacteria into two major groups: Gram-positive and Gram-negative. This technique relies on the ability of bacterial cell walls to retain a crystal violet dye. Gram-positive bacteria, with their thick peptidoglycan cell walls, retain the dye and appear purple under a microscope. On the other hand, Gram-negative bacteria, with thinner peptidoglycan layers and an additional outer membrane, do not retain the dye and appear pink after the application of a counterstain. The Gram stain is invaluable in classifying bacteria and guiding appropriate antibiotic therapy.

Another staining technique, the Acid-Fast stain, is particularly useful in identifying Mycobacterium species, including the notorious Mycobacterium tuberculosis, which causes tuberculosis. Developed by Paul Ehrlich and Friedrich Neelsen, the Acid-Fast stain offers a way to differentiate acid-fast bacteria (those that resist decolorization by acid-alcohol) from non-acid-fast bacteria.

It involves the use of carbol fuchsin, which stains acid-fast bacteria bright red, while non-acid-fast bacteria are counterstained blue. This technique is crucial in diagnosing tuberculosis and other mycobacterial infections, enabling prompt treatment and control of these diseases.

Understanding these staining techniques is vital for biology students studying microbiology, as they lay the foundation for the identification and classification of bacteria. By learning the principles and applications of the Gram stain and Acid-Fast stain, students will be equipped with essential tools for diagnosing and treating infectious diseases. Moreover, these techniques offer insights into the structural and biochemical characteristics of microorganisms, aiding in the study of microbial physiology and pathogenesis.

In conclusion, the subchapter "Staining Techniques: Gram Stain and Acid-Fast Stain" provides a comprehensive overview of these fundamental staining methods in microbiology. Aimed at students specializing in biology, this chapter offers a clear understanding of the principles, applications, and significance of the Gram stain and Acid-Fast stain techniques. By mastering these staining techniques, students will be better prepared to unravel the mysteries of the invisible microbial world and contribute to advancements in the field of microbiology.

Molecular Techniques: PCR, DNA Sequencing, and Gel Electrophoresis

In the fascinating world of microbiology, understanding the invisible world requires the use of powerful molecular techniques. In this subchapter, we will delve into three essential techniques: Polymerase Chain Reaction (PCR), DNA Sequencing, and Gel Electrophoresis. These techniques have revolutionized the field of biology, enabling scientists to unravel the mysteries hidden within the genetic material of microorganisms.

PCR, or Polymerase Chain Reaction, is a technique that allows the amplification of a specific DNA region. It is like a molecular photocopier, making millions of copies of a DNA fragment, even if it is present in trace amounts. Students can think of PCR as a molecular magnifying glass that helps scientists study the genetic material of microorganisms in greater detail. This technique has numerous applications, from diagnosing infectious diseases to identifying genetic variations in different organisms.

DNA Sequencing, on the other hand, allows scientists to read the exact order of nucleotides in a DNA molecule. It provides a comprehensive understanding of the genetic code, enabling researchers to identify genes responsible for specific traits or diseases. DNA sequencing has transformed the way we study microbiology, helping us uncover new insights into the complex world of microorganisms, their evolution, and their interactions with the environment.

Gel Electrophoresis is a technique used to separate DNA fragments based on their size and charge. It involves applying an electric field to a gel matrix, causing DNA molecules to migrate through the gel. This separation allows scientists to analyze DNA samples, compare genetic profiles, and identify genetic variations. Gel electrophoresis is an indispensable tool in forensic science, genetic research, and medical diagnostics, enabling precise identification and analysis of DNA fragments.

Understanding these molecular techniques is essential for any student studying biology. By mastering PCR, DNA sequencing, and gel electrophoresis, students can gain valuable insights into the complex world of microorganisms and their genetic makeup. These techniques have revolutionized the field of microbiology, paving the way for groundbreaking discoveries and advancements in various areas, including medicine, agriculture, and environmental science.

In conclusion, PCR, DNA sequencing, and gel electrophoresis are powerful molecular techniques that have transformed the field of biology. They have provided scientists with the tools to study the invisible world of microorganisms in unprecedented detail. By grasping the principles and applications of these techniques, students can unlock the secrets hidden within the genetic material of microorganisms, contributing to the advancement of scientific knowledge and the improvement of human lives.

Chapter 12: Microbiology in the Real World

Careers in Microbiology: Research, Medical, and Industrial Fields

Microbiology is a fascinating field that offers diverse career opportunities for students with a passion for biology. Whether you are intrigued by the study of microorganisms, interested in medical applications, or drawn to the industrial aspect of microbiology, this subchapter will shed light on the various career paths you can pursue within the realm of microbiology.

Research is at the heart of microbiology, and countless scientists dedicate their lives to uncovering the secrets of the invisible world. As a research microbiologist, you will work in laboratories, conducting experiments, analyzing data, and making groundbreaking discoveries. This field allows you to contribute to our understanding of microorganisms, their behavior, and their impact on the environment and human health. Research microbiologists often work in academic institutions, government agencies, or private research organizations.

If you have a passion for helping others and a desire to make a difference in people's lives, a career in medical microbiology might be your calling. Medical microbiologists play a crucial role in diagnosing and treating infectious diseases. They work in hospitals, clinical laboratories, and public health agencies, using their expertise to identify disease-causing microorganisms, develop effective treatments, and prevent the spread of infections. This field offers a unique combination of laboratory work and direct patient care, making it an ideal choice for those interested in both biology and healthcare.

Microbiology also finds its applications in various industrial sectors, ranging from pharmaceuticals to food and beverage production. Industrial microbiologists work in manufacturing companies, where they utilize microorganisms to produce antibiotics, enzymes, vaccines, and other products. They are involved in quality control, ensuring that the production processes are efficient and that products meet safety standards. This field requires a strong understanding of microbiology, as well as skills in biotechnology and fermentation.

In conclusion, a career in microbiology offers exciting opportunities in research, medical, and industrial fields. Whether you are driven by curiosity, a desire to help others, or a passion for innovation, there is a path for you within this invisible world. As a student of biology, you have the chance to explore the depths of microbiology and contribute to the advancement of science, healthcare, and industry. So, embrace the invisible world, and let your career in microbiology unfold.

Microbiology in Everyday Life: Food Safety and Hygiene

Food safety and hygiene play a crucial role in our daily lives, and understanding the basics of microbiology is essential to ensure the food we consume is safe and free from harmful microorganisms. In this subchapter, we will explore how microbiology relates to food safety and hygiene and why it matters to students studying biology.

Microorganisms are everywhere, including in the food we eat. While most microorganisms are harmless or even beneficial, some can cause foodborne illnesses that can range from mild discomfort to severe health issues. This is where microbiology comes into play. By understanding the types of microorganisms that can contaminate our food and the conditions that promote their growth, we can take steps to prevent foodborne illnesses.

One of the key concepts in microbiology is the concept of food spoilage. Microorganisms such as bacteria, yeasts, and molds can spoil food by breaking it down, producing unpleasant odors, flavors, and textures. By understanding the conditions under which these microorganisms thrive, we can store and handle food properly to minimize spoilage.

Additionally, microbiology helps us understand the importance of proper cooking and food preparation techniques. Heat is a powerful tool to destroy or inactivate many harmful microorganisms. By ensuring that food is cooked to the right temperature, we can eliminate the risk of foodborne illnesses.

Furthermore, hygiene practices are crucial in preventing the spread of harmful microorganisms. Proper handwashing techniques, cleaning and sanitizing food contact surfaces, and preventing cross-contamination are all essential aspects of food safety. Microbiology teaches us about the potential risks associated with poor hygiene practices and how to mitigate them.

For students studying biology, understanding microbiology in the context of food safety and hygiene is particularly significant. It provides a practical application of the theories and concepts they learn in class. It also highlights the importance of the field in real-life situations and encourages students to consider careers in microbiology or related fields.

In conclusion, microbiology is a vital tool for ensuring food safety and hygiene in our everyday lives. By understanding the role of microorganisms in food spoilage, the importance of proper cooking techniques, and the significance of hygiene practices, students studying biology can apply their knowledge to protect themselves and others from foodborne illnesses.

Microbiology and Public Health: Outbreak Investigations

In the field of biology, microbiology plays a crucial role in understanding and combating infectious diseases that pose a threat to public health. Outbreak investigations, in particular, form an essential part of microbiological research and are instrumental in preventing the spread of diseases and safeguarding communities. This subchapter explores the significance of microbiology in outbreak investigations and its implications for public health.

Outbreak investigations are conducted to identify the cause, source, and mode of transmission of infectious diseases that occur in clusters or epidemics. Microbiologists employ various techniques to analyze samples collected from patients, animals, and the environment to identify the pathogen responsible for the outbreak. They use molecular biology tools such as polymerase chain reaction (PCR) and DNA sequencing to determine the genetic identity of the microorganism and trace its source.

Understanding the epidemiology of an outbreak is crucial in controlling its spread. Microbiologists work closely with epidemiologists to collect and analyze data on the affected individuals, their demographics, and their potential exposures. This information helps determine the risk factors associated with the outbreak and aids in implementing appropriate preventive measures.

Microbiologists also play a crucial role in developing diagnostic tests for detecting and identifying pathogens quickly and accurately. These tests enable healthcare professionals to diagnose infected individuals promptly, preventing further transmission and facilitating timely treatment. Additionally, microbiologists contribute to the development of vaccines and antiviral drugs to combat emerging infectious diseases.

Public health officials rely on microbiologists' expertise to assess the risk associated with an outbreak and provide evidence-based recommendations for disease prevention and control. Microbiologists collaborate with government agencies, healthcare organizations, and research institutions to develop guidelines and policies aimed at minimizing the impact of infectious diseases on the population.

For students interested in the field of microbiology, outbreak investigations offer an exciting opportunity to apply their knowledge and skills to solve real-world problems. By studying microbiology, students can contribute to the prevention and control of infectious diseases, ultimately making a significant impact on public health.

In conclusion, microbiology plays a vital role in outbreak investigations, providing valuable insights into the causes and transmission of infectious diseases. The collaboration between microbiologists, epidemiologists, and public health officials is essential in detecting, preventing, and controlling outbreaks. For biology students, understanding the principles and techniques of microbiology in outbreak investigations opens up a world of opportunities to contribute to the field of public health and make a tangible difference in the lives of individuals and communities.

Ethical Considerations in Microbiology Research

In the field of microbiology, research plays a crucial role in advancing our understanding of the invisible world of microorganisms. However, it is important to approach this research with a strong sense of ethics and responsibility. Ethical considerations in microbiology research are essential to ensure the well-being of both humans and the environment, as well as maintaining the integrity of scientific knowledge. In this subchapter, we will explore the key ethical considerations that students studying microbiology need to be aware of.

One of the primary ethical considerations in microbiology research is the welfare of human subjects. When conducting research involving human participants, students must ensure that informed consent is obtained, and that the research is conducted in a manner that respects the autonomy, privacy, and dignity of the individuals involved. This includes protecting the confidentiality of personal information and providing appropriate compensation for any risks or inconvenience faced by the participants.

Another important ethical consideration is the humane treatment of animals used in microbiology research. Animal studies are often necessary to understand the effects of microorganisms on living organisms. However, it is essential to minimize the suffering and discomfort of animals by adhering to strict guidelines and regulations. Students must ensure that proper care and housing are provided, and that alternative methods are considered whenever possible to reduce the use of animals.

Additionally, ethical considerations extend to the environment. Microorganisms have the potential to impact ecosystems, and students must be mindful of this when conducting research. It is crucial to minimize any potential harm to the environment by implementing proper safety measures, disposing of hazardous waste appropriately, and considering the long-term effects of the research.

Integrity and honesty in scientific research are also paramount. Students must avoid plagiarism, fabrication, or manipulation of data. They should accurately report their findings and give credit to the work of others. This includes acknowledging the contributions of colleagues and collaborators, as well as disclosing any conflicts of interest that may arise.

In conclusion, ethical considerations in microbiology research are vital for students in the field of biology. By adhering to ethical guidelines, students can ensure the well-being of human subjects, the humane treatment of animals, the protection of the environment, and the integrity of scientific knowledge. Embracing ethical practices not only enhances the credibility of research but also upholds the ethical standards of the scientific community.

www.ingramcontent.com/pod-product-compliance
Lightning Source LLC
Chambersburg PA
CBHW070814170726
48000CB00017B/911